1

PHYSICS THEOREM

THEOREM

VOLUME I

Philip Mazeikas

TABLE OF CONTENTS

8

ABSOLUTE MATRICE EQUALITY

$$i^A - LQ = F$$

Coordinate value derives one.

ABSOLUTE POSITION

$$\iiint \mu^{-A}\left(\frac{GA^{11}}{-r}\right)^3 = K$$

Integer reposed beyond interval of stasis of requisite velocity defers determined axis.

ABSOLUTIA

$$\pi \sum \mu^{\int AGQ} = -t$$

Inverse coordinate function upon limit of dissension of prelocated variant pressure enters refrain.

Static pressure upon inverse function of determined limit denies function.

ADMATTER

$$\Phi^{\sqrt{j}} = L$$

Longitudinal variance upon return of absolute quotient of inert function of integral postulate defines gravitational determined axis.

AFRAME

$$\iint \Psi^{x+G} - A\sqrt{e}\left(\frac{-G}{\lambda\mu}\right) = -E$$

Dimensional inverse determinant upon instantaneous variable acceleration enters instantaneous dissolution.

AFUNCTION

$$\lambda \sqrt{\dfrac{L}{mrt}} + L = R\Omega c$$

Inverse longitudinal variance upon repose of
indetermined variable coordinate defers measure.

Postulate juncture upon inverse quotient of
absolute coordinate designates variable.

Pressure derived upon coordinate return
determines one.

ALIGNMENT

$$\pi \iint \frac{\Sigma^3}{e} = 1$$

Interposition upon return of designated value enters measure.

ANALOG

$$\boxed{\Sigma\Psi = -A}$$

Inverse coordinate force determined upon instantaneous gravitational requisite function of return defines one.

ANIMATION

$$\Sigma \iint \Psi = A$$

Interval upon repose of gravitational inverse quotient upon remand of instantaneous measure denies frequency.

ANTIAXIS

$$\iint \triangle^{\pi} \left(G \frac{-r}{a^{\lambda K}} \right)^{\sqrt{E+\Phi}} = Q - A\sqrt{f}$$

Deposition upon static juncture of instantaneous quotient upon derivative of field enters validity.

ANTIVARIABLE

$$\frac{G^3}{Z} + \sqrt{f} = F^x$$

Designated limit defines congruency.

Pressure denies instantaneous return.

Limit upon directive of pressure defines measure.

Interval upon static field defines function.

Limit denies variable.

ASCENSION

$$\Phi^{iM} - m^{\sqrt{A}} = -1$$

Interval upon juxtaposed force of intermittent gravitational pressure enters field.

Dislimit upon inert dimensional derivative of entropy defines dissension.

Juncture of interposed determinant of congruent postulate determines dimension.

Pressure upon dislocated entropy derives absolute.

BOSON

$$\int \sqrt{\lambda} - \Psi^{\sqrt{A+q}} = Q$$

Disjuncture upon harmonic integral vestige of inverse threshold denies field.

Juncture upon dissension of integral quotient enters variable.

Instantaneous congruence upon rationalized determinant of dislocation of pressure defines measure.

CALIBRATION

$$\lambda\sqrt{E}^{Lm} = -G$$

Definition upon variable function of absolute inertia designates measure.

CARRIER

$$A^G = EM^Q$$

Function upon integral measure enters force.

CAUSAL INERTIA

$$\Omega \int \frac{x}{h^y} + t\sqrt{\frac{R}{x^\Omega}} = \varphi^{P+\lambda h}$$

Variant prerequisite determinant of absolute coordinate enters intermittent function.

CENTRIFUGAL MATRICE

$$Ah\Psi - \sqrt{L} = q^A$$

Inert dissolution of interval upon designated pressure designates interval.

CHAOS

$$\iint \left(\sum \frac{-8}{lq} \right) \left[-Q \sqrt{\frac{r^{\Phi}}{mAr}} \right]^{q\pi G} = \Sigma C$$

Instantaneous quotient upon dissension of variable static juncture upon dislimit of integral longitudinal threshold enters instantaneous return.

CHEMICAL MAGNETISM

$$\Sigma\sqrt{m}[Af]^{\lambda} - \frac{\Psi}{A^{\Sigma}} = \pi$$

Insequence upon return of absolute deference upon inversion upon static field denies pressure.

Dislimit upon postulate of inert measure defines incongruent threshold of instantaneous field.

Pressure dislocated upon invariable frequency denies deference upon static limit.

COLOR

$$\Sigma\sqrt{A\lambda}^{G} + \frac{E^{\Psi}}{\pi^{A}} = q$$

Variable inertia upon dislocated measure of integral derivative of insequential pressure denies variable dissension.

Prelocated vestige upon instantaneous quotient of threshold upon dislimit of instantaneous velocity defines axis.

CONJUNCTION ABSOLUTE

$$E\mu + \sqrt{M} = K$$

Integral function upon invariable integer determines constant.

CONSTANT VALUE

$$\mu Y = L + A^x$$

Immeasure upon determined operation enters equality.

CONTINUITY

$$-EAG\left(i\Phi\sqrt{\lambda}\right)^{x+G} = -x\pi$$

Inverse longitudinal field upon determinant recourse of static variable function of derivative of prelocated measure enters invalidity.

CONVERGANCE

$$\frac{\Psi}{-E} + \left(\frac{\sqrt{Q\pi}}{\lambda^{G}} \right) = -A$$

Differential upon static quotient of indivisible
field denies incongruent axis.

Pressure upon instantaneous derivative of
postulate enters interposition.

Limit upon deference of instantaneous
gravitational inert function defines derivative.

Juncture upon deposed frequency enters inertia.

Measure designated upon invariable threshold of
velocity defines integer.

COORDINATE
PRELOCATION

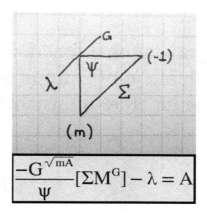

Displaced gravitational function upon inert longitudinal quotient of entropy derives static threshold upon juncture of instantaneous axis.

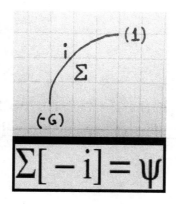

$$\Sigma[-i] = \psi$$

Postulate upon inverse indeterminant of congruent field enters acceleration.

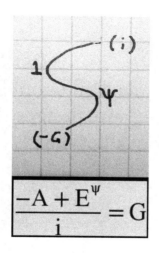

$$\frac{-A + E^{\psi}}{i} = G$$

Congruent dimension upon dissension of disjunction of measure defines derivative upon interval.

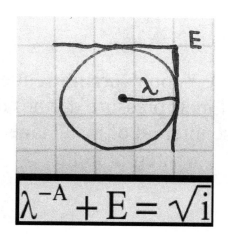

$$\lambda^{-A} + E = \sqrt{i}$$

Dimension upon variant frequency of gravitational inert pressure defines determinant upon inverse interval of instantaneous limit of integral field.

DERIVATIVE POLARITY

$$\frac{-f^G - x^3}{2}[EF]^{-1} = 0$$

Function dissension upon inverse axis defines pressure.

Limit upon variant dislocation of integral gravitational vestige of inert dimension enters congruence.

Juncture upon displaced measure enters variant threshold of instantaneous axis.

Postulate upon deferred matrice enters singularity.

DETERMINED MEASURE UPON VESTIGE OF LIMIT

Displaced frequency designation entropy inverts frequency.

$$\frac{\pi^3}{E} + \sqrt{x - E} = 1$$

The displaced inertia of determined value displaces upon the limit of designated measure.

The interval of planetary designation displaced instantaneously constant measures the interval of congruent frequency of inert limit.

$$mE - \frac{E\sqrt{x}}{\Phi} = m$$

The determined measure upon the longitudinal vertex displacing determined value of integral velocity of limit of congruent measure inverts.

Designated force beyond the limit of congruent measure of velocity of instantaneous displacement of frequency of interval of congruent limit determines frequency.

$$\sqrt{x^3} + E^\pi = \Phi$$

Limit upon frequency of interval determines measure.

DIA

$$3\lambda^r - Q = -r\sqrt{G}$$

Pressure upon dissension of invariable static
return denies variance.

DIFFERENTIAL SPHERE

$$\left(0^x - \left[G\sqrt{i} + M^\pi\right]\right) - \frac{\Phi^\kappa}{(-1)^{\left[\Sigma - \sqrt{x}\right]}} = x$$

Inexact postulate upon derived frequency denies constant.

Pressure disjunction upon field measures absolute.

Pressure upon dislocated axis derives gravity.

DIMENSIONAL PROXY

$$AE^{-i}\sqrt{G} = \pi$$

Designated measure denied upon invariable
pressure dislimits juncture.

Postulate upon inverse threshold enters
gravitational axis.

DISPERSION

$$G^{A-\Sigma}\left[\frac{A+G}{G^{\Sigma}}\right] - A^{\sqrt{G}} = 0$$

Absolute function upon inverse coordinate dissolution denies variant prerequisite field.

ELECTROMAGNETIC PULSE

$$\sqrt{A\Omega}^{i} - L = Q\varphi$$

Variable fission derived upon intermediary quotient of absolute coordinate of interposed polarity denotes calibrated function of intermediary return.

ELECTROSTASIS

$$\pi \sum \Omega^{\sqrt{G}} - [Am]^r \theta = A$$

Coordinate derivative upon repose of invalid recourse upon indivisible quotient of intermediary limit enters disjunction.

Postulate deferred upon recourse of invariable pressure enters quotient.

FISSION DIMENSION

$$\prod \varphi^{\frac{\sqrt{A\lambda}}{G}} (iM)^x = x$$

Interposition upon refrain of invariable congruent axis denies threshold.

FORCE

$$\pi^{\prod^{\sqrt{E}}} - G\Phi[A - \sqrt{i}]^7 = 1$$

Instantaneous proximate function of determined field enters variant quotient.

FORCEQUOTA

$$\Sigma\Phi = \prod i$$

Inert longitudinal constant derives stasis.

Prerequisite value deposed upon variant integer returns.

FORCEZERO

$$e\lambda - A\int q = L^t$$

Differential upon quotient of absolute interval of static dislocation of absolute determinant defines measure.

FRAMEWORK

$$M^2 \frac{AE}{\lambda} = x^2$$

Position upon dimensional absolute enters constant.

One upon determined field defines force.

Juxtaposition upon dislocated measure derives proportion.

Derivative upon absolute enters one.

Field enters variable measure.

Absolute enters dislocated force.

Matrice upon function derives one.

Measure upon absolute field enters one.

Order upon displaced measure enters force.

FUNCTION CALCULATION

$$\frac{H^r - \Psi}{\mu G} = N^H$$

Interposed variable force enters one.

Deposed integer upon incongruent field derives measure.

Analogous dislocation upon determined value enters field.

Integral measure derives coordinate.

Determined value enters product.

GRADIENT

$$\iint \prod \int (\mathrm{AM})^4 + \mathrm{E}^{\pi+\Sigma} = \mathrm{T}$$

Dislocated integer deposed upon integral variable quotient of absolute dimension of displaced gravitational axis defines intermediary limit.

GRAVITON

$$\Sigma^{i-c}\sqrt{AQ} = f^3$$

Derivative upon insequence of juxtaposed dimension of inert variant pressure defines incongruent quotient.

Postulate upon differential of static recourse and invariable juncture dislimits inertia.

Acceleration upon differential of static threshold of instantaneous dimension enters continuity.

Inert pressure upon dislocation of instantaneous constant of function of axis determines variable.

Congruent dimension upon interval of static measure derives stasis.

HYDRASPHERE

$$\frac{G^{-3}[F\sqrt{i}^{q}]}{-\Phi^{x}} = Af^{\pi}$$

Interval displaced beyond incongruent measure determines field.

Influx determined variance defines insequential determinant.

Pressure upon invariant determinant vestige upon determinant force prelocates inert dimension.

Postulate upon inertia defines absolute.

HYDRASYNTHESIS

$$\sqrt{A}^{-1} - \frac{x}{-1} = 0$$

Differential upon invariable frequency derives return.

Disjunction upon inverse threshold denies absolute.

Pressure upon gravitational measure inverts.

Postulate upon differential denies congruent interval.

Limit unjuxtaposed upon variant differential denies acceleration.

Pressure upon dislocated field denies frequency.

ILJ

$$\sqrt{r\Omega}^{A} - h = G$$

Pressure upon incongruent dissolution of integral function defines one.

IMPRESSUM

$$\lambda^H - \frac{A\sqrt{G}^\lambda}{\psi} = -G$$

Disjunction variable inertia upon dislocated matrice of instantaneous quotient of inverse field denies deference upon integral value.

Pressure dislocated upon variable function of integral measure enters instantaneous postulate upon inversion of static gravitational axis.

Postulate upon derivative of stasis enters velocity.

INCANDESCENCE

$$\sqrt{C}^{\Psi^E} - \frac{(aq)^x(-Gx)^{\sqrt{E}}}{yC} = 0$$

Deposition upon refrain of static intermediary defers instantaneous dislimit.

Intermediary axis upon deference upon interposed field denies variant dissension.

Instantaneous quotient upon refrain of intermediary proximate dimension enters value.

INFINITE FIELD

$$\pi^{\sqrt{A+\mu}}\left(\frac{G\lambda}{K^A}\right)^{\pi} = \Psi$$

Derivative influx upon instantaneous quotient of static dislimit upon determined frequency denies inversion.

INSTANTANEOUS
PROXIMITY

$$I^G - L = A$$

Deference upon coordinate designates variance.

INVALIDITY

$$\int \sum \left(-G \frac{Ac}{\Phi} \right)^{c+G} = -R$$

Derived sequential inversion upon designated
pressure enters invariable recourse.

Presupposition upon integral juncture of
determined value returns.

INVERSE SINGULARITY

Derivative upon congruent variable dissension returns.

$$-1(x + G) - \sqrt{E}^{i} = 0$$

Inanimate force denies return.

$$x^{\sqrt{G}} - i = -E$$

123

Threshold upon instantaneous limit defines variable.

$$1 - G^E = x$$

Differential upon integral variance enters limit.

$$-E^{\sqrt{i}} - G^x = -1$$

LIGHT AXIS

$$\Sigma L + \frac{AQ}{\Sigma} = M\Omega$$

Inverse variant measure upon inert disjunction of absolute inertia enters determined function.

LIGHT ORIGIN

$$A^{G+\sqrt{t}}\frac{Q}{\varphi\lambda} = \Psi$$

Inverse determined absolute threshold upon integral requisite interval of determined function derives dislocated entropy.

LIGHTFORM

$$\mu \iiint \Phi - \delta\pi = \sqrt{rh}^{a}$$

Inert variable function of interposed congruent field enters entropy.

MAGNETON

$$E^{x-L}\sqrt{r} = M$$

Interposed variant sequence of absolute
conjuncture of acceleration upon deposed
integral field denies zero.

Pressure upon disjoined function of instantaneous
derivative of interval of entropy disjoins
absolute.

Pressure deposed beyond congruence of absolute
vestige of inertia enters location.

Coordinate function of limit upon congruent
variable limit of absolute measure denies return.

MANIFORM

$$\Psi^{\iint^{\theta}} - L = Q$$

Acceleration returned upon dislocation of integral function enters conjunctive gravity.

MANUAL

Derivative upon analogous field of inert value enters force.

MATTER QUANTIFICATION

$$1 = EjR^L$$

Disjunction upon sequential function of
determined measure enters one.

MEASURE

$$-a\left(q\sqrt{i\pi}\right)^{a} = -1$$

Integer upon differential of static pressure and return upon dissolution of field denies variable.

METAVARIABLE

$$\prod \iint \mu \frac{E}{L} = a$$

Inverse longitudinal variance upon dissension of field enters coordinate.

METAWAVE

$$\Sigma \iint \delta \sqrt{\frac{Ah}{rE}} = r$$

Intermediary threshold of absolute congruence upon displaced gravitational value derives dimension.

MONISM

$$\frac{(Ac - \sqrt{A})}{-3} + i^{G} = 1$$

Juncture upon invariable field defines
instantaneous limit.

Pressure upon dissension of static variant inertia
defines congruent axis.

Instantaneous frequency upon dislocated interval
of instantaneous threshold of derivative of static
field enters integer.

MULTIPOLARITY

$$(-\pi)^m + \frac{y}{(-G)} = Q$$

Disjunction upon velocity measures static field.

Postulate upon frequency of dimensional inertia denies limit.

Gravitational field returns invariable acceleration.

Determined vestige upon disjunction of threshold measures constant.

Pressure denies frequency.

Threshold upon entropy derives one.

MULTIVARIABILITY

$$x^{(-E)}\left[\sqrt{Gm}\right] - A^x = \frac{\pi}{\Phi}$$

Denial upon variant gravitational inertia postulates constant.

Pressure upon dislocated function of dimensional threshold postulates juncture.

n

$$\frac{-Q[\Sigma a]^{11}}{r\sqrt{A}} - \Psi\left(\frac{E}{\Sigma}\right)^{n} = n^{Q^{\pi}}$$

Differential upon variant dislimit of incongruent pressure upon dislocated axis of derived threshold upon insequential variant enters intermediary field.

Prelocation upon influx of static designated interval upon congruence enters absolute.

NEGATIVE CONSTANT

$$\pi\sqrt{p\Omega}^{\,j} = r$$

Coordinate function upon inert longitudinal measure denies determined limit.

NEUTRALITY

$$\pi^3 = \frac{K^x}{Mq}$$

Differential upon instantaneous quotient of interposition of field denies inverse determined axis.

NEXUS

$$\frac{\lambda\sqrt{Ac^m}}{-r} = \Sigma\sqrt{x}$$

Interval upon instantaneous congruence of inert static dislocation of invariable threshold denies frequency.

Postulate upon derivative of juncture of inverse determined limit denies force.

OMDIMENSION

$$\lambda \frac{-2}{Ay}[\sqrt{H}]^{G-Q} = M$$

Coordinate velocity upon repose of dissension of static interval upon inverse axis of interposition of field enters measure.

OMNIAXIS

$$\frac{-1}{i^E} + G\sqrt{\psi} = \Sigma$$

Instantaneous measure upon inverse determinant of prelocated field derives pressure.

Inert value upon interval of instantaneous juncture enters differential of static measure and field.

Postulate upon derivative of gravitational definition of threshold of absolute velocity enters zero.

Quotient upon differential of inert pressure and prelocated dissension defines congruence.

Inert return upon interval of static dimension defines absolute inertia.

ONE

$$Q^{\frac{L}{Hr}} - \frac{L}{Q}(Aw)^{\sqrt{i+G}} = 0$$

Dissension upon field of incongruent variable quotient of stasis enters determined integer.

ORBITAL AXIS

$$\sqrt{A\mu}^{\,i-A} = G$$

Interposition upon refrain of static derived postulate defers incongruence.

OVERVERSE

$$A\Phi\sqrt{H} = 1$$

Dimension upon designated value enters function.

PHI

$$\Sigma\left(\pi + \frac{-x}{\sqrt{\Sigma}}\right) - \frac{a}{-\pi^{\Psi\lambda\pi}} = \Phi^x$$

Dislimit upon interposition of refrain upon dislocated function of inverse dissolution upon threshold of field enters rational prelocation of instantaneous quotient.

PHOTON HARMONIC

$$\sqrt{Af\pi^i} = E$$

Dislimit upon gravitational function of derived
pressure enters variable.

Prelocated variable integer upon instantaneous
congruence dislimits postulate.

Interval upon gravitational threshold of inert
measure defines field.

PHOTONIC DIVISION

$$\Phi^{\sqrt{\mu}}\frac{\Omega}{i} = A$$

Intermittent longitudinal variable sequence of interposed coordinate magnetism derives instantaneous axis.

PIQUE

$$-\Phi\prod[G\Psi]^A = a^A$$

Dissension upon polarity derives incongruent displaced variance.

Integral value deferred upon recourse enters intermittent stasis.

PITE

$$\Omega\sum RQ = m$$

Analogous derivative upon determined wave function of absolute limit denies coordinate.

PLASMASYNTHESIS

$$\left(\sqrt{m} - V\right)^2 = A$$

Instantaneous return derives interval.

Measure upon disjunction defines pressure.

Derivative upon gravitational derivative denies frequency.

Velocity upon return inverts.

Instantaneous field measures static interval.

Determined threshold defines instantaneous postulate.

PREFLUX

$$\Sigma^{\Omega - A} = m$$

Instantaneous interval upon refrain upon intermittent value defers postulate.

PREFUNCTION

$$\psi^{\sqrt{E}} - G = A$$

Dissolution upon field enters prerequisite measure.

PREMEASURE

$$R^a - \lambda = LQ\mu$$

Interval upon axis of dislocated differential of disjunction of gravitational function and determined entropy derives one.

QLA

$$\lambda \frac{\Phi^{i-a}}{M^3} = E$$

Interposition upon refrain of absolute measure enters stasis.

QUADRATIC FORCE

$$\theta \prod i[qh] + \varphi = M^t$$

Intermediary threshold upon variant degree of acceleration denotes coordinate.

Presupposition of field upon denotation of postulate defers congruence.

Pressure upon intermediary value defines constant.

Proximate return upon variant integral field enters requisite function.

SINGULAR DESIGNATION

$$\Sigma \prod h \sqrt{Lm} = m$$

Entropy derived upon instantaneous measure of coordinate limit upon inverse determined function enters inert disjunction.

SINGULAR LONGITUDE

$$\mu\prod\sqrt{\Sigma}^{A-q} = 1$$

Derivative upon repose of instantaneous
threshold of disjunction upon field enters
absolute.

Postulate upon return upon instantaneous
measure denies field.

SINGULARITY

$$L^{mq}\sqrt{A} = A$$

Determined function upon determined value
enters one.

SQUARE QUOTIENT

$$\left(g^{\lambda} + \Sigma F\right) - \sqrt{Q} = C$$

Instantaneous interval upon dislocated variant field defines congruent dimension.

Postulate upon integral value derives static function of instantaneous threshold of acceleration.

Pressure upon displaced variant axis defines instantaneous measure.

STABILITY

$$\iiint \varphi - H^{\sqrt{A}} = Q$$

Inert quadratic dimension of absolute inertia denies constant.

STASIS DISJUNCTION

$$EC = M\pi^E$$

Pressure integral upon absolute measure enters one.

Value designated upon constant function defines interposition.

One enters function upon immeasure.

STASIS INTERPOSITION

$$\mu^i = Qx^q$$

Repose upon variant dissolution of prerequisite singularity of measure denotes variable absolute.

SUBSONIC

$$\prod\iiint \lambda - A^x \sqrt{3} = -A$$

Interposition upon dissemination of field upon inverse determined threshold of instantaneous proximate axis dislocates juncture upon instantaneous analogous return.

SUSPENSION

$$\Omega \sum \int K - \sqrt{Lr}^{AQ} = -A$$

Intermittent analogous return upon variant dissension of coordinate velocity determines absolute.

THE HELIX

$$-A^{\Phi} + \frac{\left(\sqrt{\pi}\right)^i - E}{G^7} = A$$

Postulate upon differential of inverse axis derives instantaneous limit.

Pressure upon gravitational field defines inert variance.

Dislocated velocity upon differential of static velocity derives instantaneous axis.

Postulate upon derivative of static measure enters field.

THE HELM

$$\sqrt{g}^{3} + i = R$$

Dissension upon field of inverse determined constant defers instantaneous quotient of derivative of pressure.

Inert threshold upon instantaneous derivative of static interval defers interposition.

Field upon dissension of static dimension derives inert value.

THE INDETERMINANT

$$G^{\lambda}\sqrt{AE} - E(x,\psi) = -x$$

Dissolution upon refrain of dislocated function of variance defines conferred force.

Pressure upon interval of determined variable sequence enters threshold.

Instantaneous quotient of static gravitational juncture upon derivative of inert axis enters zero.

THE LEVEL

$$\sqrt{E - q} + Q = A$$

Presupposition upon interval of coordinate limit enters value.

THE NUCLEAR FORCE

$$\frac{\Sigma^{\sqrt{A}} + G^4}{-i(\lambda - \Phi)^r} - \pi^{(-E)} = Z^\varphi - \triangle\left(\sqrt{gA}\right)^i$$

Disjunction variable function defines instantaneous entropy.

Derivative upon threshold inverts.

Function defines dissension.

Pressure upon velocity of measure determines zero.

Instantaneous return upon congruent variable field derives static gravitational acceleration.

Displaced force upon invariable threshold denies variable.

THE PARTICLE

$$\frac{[oQ]^a}{-mq} + \frac{x^4 M^C}{\sqrt{a}} = -Q$$

Constant instantaneous inertia disjunction derives
invariable fission.

Limit dislocated upon variant field enters
derivative.

Pressure upon inert variable dissension returns.

Congruent determined function of proximate
frequency denies variance.

Limit dislocated upon inert value derives stasis.

THE SQUARE ORIGIN

$$\lambda \sum \Omega^8 = L$$

Determinant recourse upon invalid deference of interval of coordinate derives axis.

TIME AS FORCE

$$\Psi K - \triangle G \frac{A\Psi}{E} = L$$

Differential upon variant field enters requisite analogous prelocated axis upon deferred measure.

Interanalogous requisite function upon derivative of frequency denies coordinate juxtaposition upon stasis.

TIME MATRICE

$$E - q2t[Am]^E = r$$

Differential upon repose of invariable congruence upon dimension of static longitudinal field derives coordinate.

TIME PRIMARY

$$\Psi = A\mu \frac{i}{\Psi}$$

Coordinate value instantaneously measures determinant.

TORRENT

$$\iiint \theta \frac{Li}{r}[Q\Phi]^x = A\Sigma h$$

Constant inertia variable upon intermittent axis of designated threshold of derived integral measure defines velocity.

TOTAL EQUILIBRIUM

$$\sum \mu^{\sqrt{A}} - \frac{D^{\Omega}}{V\sqrt{Ax}} = 0$$

Absolute recourse upon indivisible function of acceleration upon variable field denies entropy.

TOTAL EXPENDITURE

$$\pi^H \frac{HL}{\Phi^A} = Q\Sigma^L$$

Function upon determined limit defines variable inertia.

Coordinate upon return of absolute force enters juncture.

Measure upon determined threshold of quantification of function determines limit.

Force upon dislocated axis of integral measure denies function.

Limit upon variable denotation of one enters entropy.

TOTALITY

$$\frac{-i}{r} - M\sqrt{i^x} = x$$

Postulate upon inverse variance of stasis enters pressure.

Sequential derivative upon interval of intermittent variable function of inert dislocated field defines one.

TRAKE

$$\pi\varphi - L^{A\Phi}\sqrt{EL}^{R+\varphi} = \pi$$

Inverse determined variable upon quantitative return of inordinate field upon coordinate stasis defers measure.

TRANSINTERNAL

$$L^{MG} = A\Omega M$$

Operative limit upon disjoined measure enters one.

Determined magnitude of function determines value.

Disjunction upon intermediary force determines absolute.

TRIBUTARY

$$K\int -A\left(G^A + \sqrt{E}\right)^{i-r} = -E$$

Disposition upon refrain of intermediary analogous threshold upon inverse determined sequential variable enters intermittent dislocated postulate.

UNITY

$$\sqrt{y} - K^{\sqrt{y}} = y$$

Inexact variant postulate upon integral return upon instantaneous pressure enters dislocated integral velocity.

UNIVERSAL AXIS

$$\Omega \prod \iiint P^L - A^E = R$$

Determined disjunction upon return of absolute field enters force.

VINERTIA

$$\mu^A + \Phi \frac{L}{a} = G^\mu$$

Intermediary value deferred upon congruent limit of derived pressure enters constant.

WAVE FUNCTION

$$-\sqrt{A^{-mV}}[iA]^{Gx} = -oM$$

Rationalized inderivative prelocation longitudinal proximate analogous gravitational inertia enters one.

Proximate definition upon derivative function dislocates variance upon integral axis.

Juncture upon invariable field dislimits frequency.

X-1

$$\sqrt{a\Sigma - q}^{K} + \frac{i^{G-x}}{x\sqrt{G+a}} = K^{\sqrt{\Sigma+x}} - \frac{aQ}{x^i}$$

Derivative upon proximate analogous inverse field of static entropy enters instantaneous disjunction of invariable derivative of continuity.

Prelocated variance upon inert static return of inverse determined postulate upon derived sequential limit defines interval.

Presupposition upon inert dislocation of variant integer defers quotient upon interposition of deferred dislocation of proximate axis.

CPSIA information can be obtained
at www.ICGtesting.com
Printed in the USA
BVHW051459240720
584413BV00005B/185